W9-BID-827

Oil Spills

Other Save-the-Earth Books by Laurence Pringle

Living Treasure:
Saving Earth's Threatened Biodiversity

A SAVE-THE-EARTH BOOK

LAURENCE PRINGLE

Morrow Junior Books · New York

Permission for the following photographs is gratefully acknowledged: AP/Worldwide Photos, pp. 8, 17, 18, 24, 29, 30, 46; Alaska Fish and Game Department, pp. 14, 34 (both), 36; American Petroleum Institute, p. 7; Exxon Corporation USA, pp. 5, 16, 35, 44; George C. Page Museum, p. 2; Pennsylvania Historical and Museum Commission Drake Well Museum collection, p. 3; U.S. Coast Guard, pp. 22, 39, 41, 43; U.S. Fish and Wildlife Service, p. 31; Visuals Unlimited, p. 33. All other photographs by the author. The chart on page 13 was excerpted from *The Oil Spill Intelligence Report*.

Design by Trish Parcell Watts

Printed in the United States of America.

1 2 3 4 5 6 7 8 9 10

Library of Congress Cataloging-in-Publication Data

Pringle, Laurence P.

Oil spills / by Laurence Pringle ; illustrated with photographs.

p. cm.

Includes bibliographical references and index.

Summary: Describes petroleum and its uses, examines the harmful effects of oil spills, and discusses how such environmental disasters can be cleaned up or prevented.

ISBN 0-688-09860-6.—ISBN 0-688-09861-4 (lib. bdg.)

1. Oil spills—Environmental aspects—Juvenile literature. 2. Petroleum—Juvenile literature. [1. Oil spills. 2. Petroleum. 3. Environmental protection.] I. Title. TD427.P4P74 1993 363.73′82—dc20

92-30348 CIP AC

This book is printed on 100 percent recycled paper.

Contents

Introduction

Oil has been called black gold. It lubricates engines and heats homes and businesses. Made into gasoline, it fuels nearly a billion automobiles, trucks, and other vehicles all over the world. Oil is found in hundreds of other products—plastics, detergents, clothing—that touch our lives every day. Oil also grows more precious every day, because its supply on earth is limited and dwindling.

Whenever oil is drilled from the earth's crust, transported, or used, it may be spilled. An oil spill is often a double disaster: a valuable resource is wasted and a bay, beach, or other natural habitat is fouled by the dark and sometimes deadly liquid.

Oil spills can kill. The 1989 *Exxon Valdez* spill of 258,000 barrels of oil killed an estimated 100,000 birds and several thousand sea otters in Alaska's Prince William Sound. And yet this spill ranked just twenty-ninth among the world's biggest oil spills. As more and more oil is shipped around the world, the threat of major oil spills grows.

This book gives basic information about petroleum and its uses, and the many ways in which both large and small amounts are spilled. It describes what happens to oil after it is spilled, and what we have learned about oil's sometimes-disastrous effects on plants and wildlife. Finally, it describes the struggle to clean up oil spills, and tells what steps must be taken to prevent more of these double disasters.

1

Precious Petroleum

Crossing a rain-slick street, you suddenly see a flash of color at your feet. It is an oil spill. A few drops of oil from an automobile have spread over the wet pavement and reflect a rainbow of colors.

Perhaps early humans first noticed oil in a similar way, as it colored the waters of a swamp or other wetland. A British dictionary, published in 1846, defined petroleum as a black liquid, "floating on the water of springs." The formal name of oil is petroleum, from the Latin words *petra*, "rock," and *oleum*, "oil."

Oil in one form or another sometimes rises naturally to the earth's surface. In 1992, following earthquakes, oil began seeping from the Santa Susana Mountains northwest of Los Angeles, California. The world's most famous oil seepage site is also in California: the Rancho La Brea tar pits of West Los Angeles. The

Use of oil has great impact on the landscape, and on our everyday lives.

sticky black tar that long ago trapped saber-toothed cats, dire wolves, and giant ground sloths is a form of petroleum. Also called asphalt, the tar has seeped from beneath the Los Angeles area for the past 35,000 years. Today, pigeons and squirrels sometimes meet the same fate as long-extinct mammals: They become mired and die, and their bones are preserved in the sticky asphalt.

Most scientists believe that the earth's petroleum formed millions of years ago, when large areas of today's continents were covered by warm shallow seas. Life thrived in these waters and on the sea bottom. As the abundant plants and animals reproduced and died, their remains fell to the sea floor.

Covered by mud and other sediments, this organic material began to change. Pressure from the sediments transformed it into compounds of hydrogen and carbon, called hydrocarbons. A rich variety of hydrocarbons makes up petroleum. (Because petroleum, natural gas, and coal all formed from organisms that existed millions of years ago, they are called fossil fuels.)

Now-extinct species of mammals were trapped in tar, and their bones were found at Rancho La Brea.

The same forces that changed plant and animal remains into fossil fuels also changed the sediments surrounding them into rock. There are about seventy major regions of sedimentary rocks in the United States. These are the regions where geologists hunt for oil.

Sometimes a fault, or crack, in the rocks allows petroleum to seep to the surface; but more often, oil is trapped underground, beneath layers of salt, limestone, or shale. It does not lie in underground pools but within the spaces of porous sandstone rock. Sandstone formations may be several miles long and hundreds of feet thick. They act like a giant sponge for oil.

A typical oil deposit includes natural gas and salt water. Both the gas and oil are lighter than water and rise to the top of the underground deposit. When a drill cuts into the sandstone, natural gas rushes out first, followed by oil. When an oil well begins to produce salt water, this usually means it has exhausted most of the easily recovered oil below.

Such typical underground reservoirs of lightweight oil are just one source of petroleum. Some oil deposits include extra-heavy oil, which is denser than water but still fluid. Venezuela has the world's largest stores of extra-heavy oil. Western Canada also has large deposits of extra-heavy oil and of tar sands, from which a type of oil called bitumen can be produced. (Bitumen must go through a process to be upgraded to good-quality petroleum.) The Western United States, Canada, and other nations also have oil shales from which petroleum can be obtained.

The first oil well in the United States was drilled by a steam-powered bit in 1859, near Titusville, Pennsylvania. In that area, oil seeped from deposits that lay close to the surface. The drill struck oil just seventy feet deep. Today, drillers must often reach down two miles in their quest for oil.

Many people believe that humans first used oil when the automobile was invented, but human use of petroleum has been traced back six thousand years. People used tar or asphalt as a mortar to hold materials together in walls

Nineteenth-century well-drillers found oil deposits close to the surface in western Pennsylvania.

and roads, and even to embalm mummies and to hold jewels in their settings. Liquid petroleum was thought to have medicinal value and was also used in lamps.

Oil could be made to burn fiercely, so it became a chemical warfare weapon. Petroleum-filled trenches, set aflame, were used to defend cities in ancient times. Oil was also an ingredient of Greek fire, a chemical weapon invented in the year 660 A.D. Greek fire was used against Arabian ships that attacked Constantinople in 673. At close range, Greek galleys fired jets of this fiery liquid through tubes onto the Arab galleys. Water thrown on the flames had little effect and the Arabian fleet was nearly destroyed.

In 1498, Christopher Columbus discovered petroleum, in the form of a large tar lake, on the island of Trinidad. He and others used the tar to fill cracks in their ships' hulls and keep them waterproof.

Liquid petroleum is rarely used so directly today. It is called crude oil, and must be altered chemically (refined) for its many uses. People have tinkered with its complex hydrocarbons for centuries. In 1852, Polish farmers asked a pharmacist to see if he could distill vodka from the oil that seeped into their fields. He boiled petroleum, then collected and cooled the vapor that was given off. The result was not vodka but kerosene, which proved to be a low-smoke fuel for lamps of that time. It replaced whale oil as the common lamp fuel.

Kerosene is one component, or fraction, distilled from petroleum. Boiling crude oil at different temperatures produces other fractions. For example,

At a refinery, oil is separated into its chemical fractions.

crude oil boiled at a temperature of 215°F yields gasoline. Other fractions include benzene, fuel oils, lubricating oils, paraffin, and petrolatum (also called petroleum jelly).

Each fraction can be refined further into other products. The residue left after oil is distilled is also useful. It contains most of the resin and asphalt fractions in petroleum. These can be used as fuel oil or for paving.

Some of the less valuable petroleum fractions can be converted to valuable products. In the conversion process, called cracking, heavy oils are exposed to high temperature and pressure. The heavy molecules break down and recombine as gasoline or other lightweight fractions. Conversion processes give us acetylene, ammonia, synthetic rubber, explosives, plastics, fertilizers, and many other materials. The distilling and cracking of oil at more than nine hundred oil refineries worldwide yields chemical compounds that are part of an estimated three thousand products. Some are trivial. Some are vital.

We live in a petroleum- or hydrocarbon-based society. Oil has altered the landscape: the petroleum fractions that power and lubricate automobiles have created suburban sprawl, modern highways, and parking lots, too.

Oil is also part of our everyday lives, at work and play. Imagine trying to get through a day without using any petroleum-based materials. This rules out most modern transportation. It may leave you without electricity, if you live where electric power is generated by burning oil. (In all power plants, of course, oil lubricates moving parts.)

Giving up petroleum may eliminate some of your clothes and footwear, furniture, appliances, games, and food. (Petroleum fractions are used in artificial fertilizers, and as a fuel and lubricant for farm machinery and for the trucks and other vehicles that transport the food.) You would also have to put off some reading, since many printers' inks contain petroleum. This book, however, is printed with a soy-based ink.

Our hydrocarbon-based society uses a lot of oil. In 1990, the United States consumed about sixteen million barrels of oil each day. (A barrel contains forty-two gallons of oil.) About half of this petroleum was imported. Worldwide, about sixty million barrels of oil are used daily.

Nearly three million wells have been drilled in the United States to help satisfy the nation's hunger for oil and natural gas. Most of the large oil fields on land have been discovered already. Trying to satisfy the world's huge appetite for petroleum products, the oil industry builds platforms offshore and probes the sedimentary rocks of the ocean floor.

Oil is also moved long distances through pipelines. Today, there are more miles of oil pipelines than of railroads. To transport oil at sea as cheaply as possible, oil companies build huge ships, called supertankers. Supertankers are more than three football fields long. They carry many millions of gallons of oil economically, but are hard to maneuver. It takes them several minutes, and several miles, to turn. Oil tankers sometimes collide with rocks or with each other.

In all of these activities, and in the everyday use of petroleum products, oil is spilled.

Each day the 800-mile Trans-Alaska Pipeline carries as much as two million gallons of oil to the port of Valdez.

2

Here a Spill, There a Spill

In 1989, a 987-foot-long oil tanker, the *Exxon Valdez* (pronounced Val-DEES), slammed into a reef off the coast of Alaska. Nearly eleven million gallons of crude oil poured through gashes in the ship's hull into the clear waters of Prince William Sound. The oil blackened twelve hundred miles of shoreline.

This was the largest oil spill ever in United States waters. Television and other news media reported on the catastrophe for many days. The name *Exxon Valdez* was etched in the memories of many young people. It was the first big oil spill of their lives.

Each generation remembers certain oil spills. Some older people remember the *Torrey Canyon* spill of 1967, in which 119,000 tons of oil leaked into the English Channel. Many people in the United States still link the words "oil spill" with the name Santa Barbara. In early 1969, the drilling of an offshore

In 1969, workers spread straw to absorb oil that washed onto beaches at Santa Barbara, California.

oil well near that city caused a leak that fouled about a hundred miles of California coastline.

A list of some of the world's largest oil spills appears below. But each spill is different, and such a scorecard doesn't begin to tell the story of these events, or of all the petroleum that enters the environment. An authoritative study, *Oil in the Sea*, was published in 1985 by the National Academy of Sciences. It estimated all sources of oil pollution in the ocean, including natural ones.

THE BIGGEST OIL SPILLS

Nine of the largest oil spills, given in order of amount spilled, are listed below. The amount of oil spilled is given in metric tons, each of which is about seven barrels. The amount of spilled oil may not be a measure of the harm done; that depends on the location of the spill and other factors. For every huge oil spill there are thousands of smaller ones. The *Exxon Valdez* spill of 1989, for example, was "only" about 37,000 tons.

Name and place	year	cause	tons
Ixtoc-I oil well, Gulf of Mexico	1979	blowout	475,000–600,000
Atlantic Express and *Aegean Captain*, off Trinidad and Tobago	1979	collision	300,000
Castillo de Bellver, off S. Africa	1983	fire	250,000
Amoco Cadiz, off northwest France	1978	grounding	223,000
Torrey Canyon, off southwest England	1967	grounding	119,000
Sea Star, Gulf of Oman	1972	collision	115,000
Urquiola, off northern Spain	1976	grounding	100,000
Hawaiian Patriot, northern Pacific	1977	fire	99,000
Braer, Shetland Islands	1993	grounding	85,000

Yes, nature itself spills oil—an estimated 200,000 metric tons each year. A quarter of this oil escapes from oil-bearing rocks as they erode. The rest spews from cracks in the sea floor. Earthquakes have fractured the sedimentary rocks of the continental shelves all along the western coast of North America. More than two hundred natural seeps have been located. They include fifty-four offshore of southern California, and more than thirty off Alaska. Each year, as much as thirty thousand metric tons of oil are released in one area north of Santa Barbara. Tar also seeps from the sea floor there.

Gallons, Barrels, Tons

Amounts of oil produced and used are expressed in either gallons or barrels. There are forty-two gallons in a barrel. In shipping, oil is usually measured by the metric ton. (A metric ton is a thousand kilograms [2204 pounds], as opposed to the English measure 2000-pound ton.) The exact number of barrels in a metric ton varies because the weight of different kinds of petroleum varies. In general, seven barrels of oil weigh one metric ton.

Oil also reaches the oceans as a result of a variety of human activity on land. There are spills and leaks at oil refineries; oil is discarded with industrial wastes; and at service stations, homes, and on the streets. People spill gasoline, engine oil, and other petroleum products every day. The individual amounts spilled are small, but they add up to many tons.

The U.S. Environmental Protection Agency estimates that some 175 million gallons of used oil are poured down drains or in landfills by people who change their own car engine oil. Some of this oil reaches rivers, lakes, and the

seas. So do some of the petroleum products that are found in municipal waste water. Sewage treatment plants release 200,000 metric tons of petroleum each year, according to the National Academy of Sciences. These plants discharge twice as much oil into coastal waters as tanker accidents.

Another often overlooked source of oil pollution is underground spills of petroleum products. There are small ones, from gasoline station storage tanks, and big ones, from refineries and large storage tanks. Over four decades (1940s–1970s), an estimated seventeen million gallons of petroleum (including heating oil, kerosene, and gasoline) leaked from storage tanks and pipelines in Brooklyn, New York. It contaminated groundwater and creeks, and spread under residential neighborhoods. One of the companies responsible for the spill, the Mobil Oil Corporation, will spend tens of millions of dollars in an attempt to pump the oil from the ground. The process is expected to take several years.

Used oil drained from vehicle engines is often discarded into the environment.

Dramatic, large oil spills at sea attract international attention, but every year there are thousands of small spills. The U.S. Coast Guard keeps a tally, and in most years counts over ten thousand incidents of oil spillage in waters of the United States alone. A sample of these incidents from the year 1990 is listed below.

"LITTLE" SPILLS

Every year there are thousands of oil spills that may not create headlines, but which waste petroleum and harm the environment. Here is a sampling of some (each of at least 10,000 gallons) that occurred in 1990.

location	gallons	cause
Wakasa Bay, Japan	237,000	tanker split in two
Freeport, Pennsylvania	75,000	pipeline rupture
Oahu, Hawaii	16,800	collision with buoy
Burrard Inlet, Canada	10,000	collision with barge
Manchester, Washington	47,300	storage tank leak
Minnesota River, Minnesota	75,000	pipeline malfunction
Tenerife, Canary Islands	10,000	tanker struck dock
Suez Canal, Egypt	2,000,000	tanker ran aground
Pacific Ocean, Japan	248,700	tanker sank in typhoon
Perth Amboy, New Jersey	100,000	storage tank collapse
Hudson River, New York	164,000	tanker hit reef
Lake Charles, Louisiana	21,000	pipeline weld failure
Corpus Christi, Texas	12,600	leak in tanker
Toledo, Colombia	294,000	storage tank bombed by guerrillas

According to the National Academy of Sciences, however, accidental spills account for only one-twentieth of the total oil pollution that occurs during oil transportation. Routine steps taken during the transporting of petroleum dumps much more oil into the environment.

Some oil is spilled during lightering—the process of pumping oil from one ship to another. Today's supertankers are too big to dock at many ports, so their oil must be lightered to smaller vessels and barges, which then carry the petroleum to shore. Oil spilled during lightering is usually a result of mechanical problems or human error.

Oil was lightered from the stricken Exxon Valdez *into another tanker. Fortunately, only about a fifth of the huge supertanker's cargo was spilled.*

Tankers also release petroleum into the environment during routine cleaning of their holds. More than one percent of a tanker's cargo usually remains behind when the ship is unloaded. To rinse out a ship's hold, the crew is supposed to pump in seawater, then pump it into special tanks on-board, where most of the oil separates from the water. The oil can then be added to the next cargo. The oil-tainted seawater is dumped into the ocean. International agreements limit the percentage of oil released, but these rules are often broken.

The tar fractions of petroleum tend to stick to the cargo compartment walls of oil tankers. Tar is frequently washed—illegally—from tankers at sea. As oil-tanker traffic across the seas has grown, so have reports of tar masses washing up on unpolluted areas, such as the east coast of Africa, Bermuda, and many islands in both the Atlantic and Indian oceans.

Petroleum is also spilled into the ocean during routine operations of offshore oil rigs. At sea or on land, oil that is pumped to the surface is usually mixed with water. This is known as produced water, the amount of which varies from well to well. Sometimes this water is pumped underground, back into the oil reservoir from which it came. Sometimes, however, produced water is discharged into the environment after going through a process that removes most of the petroleum. Nevertheless, some oil remains in the water. The National Academy of Sciences estimated that the produced water discharged annually in the United States and off its coasts carries between fifteen hundred and three thousand metric tons of oil.

The worst accidental oil spill of any kind occurred at an offshore drilling site in the Gulf of Mexico, fifty miles east of Mexico. On June 3, 1979, an exploratory well blew out and began spewing oil into the gulf waters. By the time the Ixtoc-I well was capped, 290 days later, about 475,000 metric tons of petroleum had been released into the environment. Since the spill occurred far offshore, most of this oil did not foul beaches.

Overall, offshore drilling platforms have not been a major oil spill problem. The Ixtoc-I spill occurred as a result of human error—the cause of countless spills, large and small. Mechanical breakdowns also cause many spills. Oil tankers sometimes break apart at sea. In 1978, the steering mechanism of the *Amoco Cadiz* failed. The tanker ran aground off the coast of France, split in half, and spilled 223,000 tons of oil, which contaminated nearly two hundred miles of coastline.

Many small oil spills are deliberate. They may occur, as mentioned earlier, when a tanker's crew discharges tank-cleaning water into the sea. Warfare may bring another kind of deliberate spill, as tankers, refineries, storage tanks, pipelines, and offshore drilling platforms become targets for destruction. Oil production facilities were attacked during the eight-year (1980–1988) war between Iran and Iraq. In 1983, Iraqi rockets destroyed several Iranian offshore oil rigs, causing hundreds of tons of oil to gush daily into the Persian Gulf. These wells were not capped for several months.

The worst accidental oil spill occurred in 1979 as an offshore well was being drilled in the Gulf of Mexico.

During the Desert Storm conflict of 1991, Iraq sabotaged oil facilities in Kuwait. It dumped the cargoes of several tankers, blasted onshore oil storage tanks, and released oil from the Sea Island loading area, spilling an estimated six million barrels into the Persian Gulf. On land, Iraqi troops set 752 oil wells ablaze. The last well fire was extinguished in November 1991, but about two hundred oil lakes remained. Some were more than a mile long and three feet deep. The Kuwaiti government struggled for several years to clean up this new kind of human-made disaster.

Clearly, huge amounts of petroleum spill into the environment all over the world. Even wild, isolated Antarctica is not safe; in 1989, a supply ship ran aground there and spilled 170,000 gallons of jet and diesel fuel. Some petroleum seeps naturally from the earth's crust, some gushes from stricken ships, and even more flows and drips from countless small sources.

What happens to this petroleum, and what effects do oil spills have? What we know so far is described in the next chapter.

Iraq deliberately spilled oil from Kuwait's Sea Island Terminal in 1991.

3
After the Spill

News media and the general public view a major oil spill as a disaster for sea life. The oil industry has a different view. Here is how it was expressed in a 1977 issue of *Exxon Today*: "Though manmade oil spills are unattractive and wasteful, they are not inevitably harmful to life in the sea . . . Most spills cause little damage to biological elements of the oceans."

The truth lies somewhere between these views. Spilled oil *can* cause great damage, but what effects oil has when it is spilled depend on many factors. Harm to living things depends not only on the amount of petroleum spilled, but also on what kind is spilled. For example, is it crude petroleum or gasoline? It also depends on where and when a spill occurs; and on the season, weather, and many other conditions after it happens.

The differing effects of two small oil spills along the coast of Maine illustrate

In 1993, the Braer *spilled 85,000 tons of crude oil off the Shetland Islands. Powerful winds and waves quickly dispersed the oil.*

this point. In 1971, five thousand gallons of jet fuel and heating oil leaked from storage tanks on land and leached into Long Cove at Searsport, Maine. This was a small, underground, invisible oil spill. It was discovered because it killed a population of soft-shell clams that had thrived in Long Cove.

In the summer of 1972, the *Tamano*, a Norwegian tanker, grazed an underwater ledge and spilled forty thousand gallons of fuel oil into Casco Bay. Thousands of boats had to be cleaned, sea plants and clams died, and the state closed contaminated areas to clam digging.

A decade later, Dr. Edward Gilfillan of Bowdoin College compared the two

---

Hundreds of Hydrocarbons

Large or small, each spill of petroleum is an individual event, partly because crude oil and its refined products vary so much from each other. In addition, crude oil from one field differs from that of another. Indeed, oil from one well may differ from that of another well that taps the same underground reservoir. (Chemists can tell one from the other by a process they call "fingerprinting." Many times, this process has been useful in identifying the source of a spill, and fixing legal blame for it.)

Petroleum's great value has led chemists to study it intensively, and to understand it well. Hundreds of hydrocarbon compounds make up from fifty to ninety-eight percent of crude oil. Varying amounts of sulphur, oxygen, nitrogen, and such metals as nickel make up the rest. The most abundant compounds are called alkanes. Another group of hydrocarbons, called cycloparaffins or naphthenes, sometimes makes up half of an oil sample. A third category, called the aromatic hydrocarbons, may total twenty percent of all hydrocarbons in some crude oils. The aromatics include benzene, toluene, phenanthrene, and xylene—all poisonous or toxic. Crude oil also contains small amounts of resins and asphaltenes.

---

sites. He said, "At Casco Bay now, you'll have to look hard for signs of the *Tamano*. Visibly and biologically, there has been excellent recovery.

"In Searsport, there has been practically no recovery. With that clay sediment in Long Cove holding that oil in place, there has hardly been any change in almost ten years. The clam population just hasn't come back." But eventually, like a wound healing, Long Cove did recover.

Even though oil spills at sea vary greatly, the study of several spills has given us a general picture of what happens when crude oil is released. "Weathering" is the term that scientists use to describe how an oil spill is affected by winds, waves, sunlight, and other factors in the environment. Weathering changes oil both physically and chemically.

Oil is lighter than water, so spilled oil at first stays at, or rises to, the surface. Then it begins to spread—the first step in the weathering process. The speed and pattern of its spread depends on the water temperature, currents, and the direction and speed of winds. Of course, the amount of oil spilled also affects how far oil spreads over the surface. In 1976, the *Argo Merchant* dumped nearly eight million gallons of heavy industrial oil off Nantucket, Massachusetts. At first, this spill formed a thick slick covering just a square mile. Eventually, it spread out in a thin, oily sheen that covered more than eleven thousand square miles.

As soon as spilled oil is exposed to air, parts of it begin to evaporate. The most lightweight compounds of petroleum, which include benzene and other poisonous hydrocarbons, evaporate first. They usually enter the atmosphere within a few hours of a spill. As the process of evaporation continues, slightly heavier fractions enter the air. Over a span of one to two weeks, as much as half of an oil spill may escape into the air. (Spilled gasoline, jet fuel, or kerosene may evaporate completely.) As evaporation removes the lightest hydrocarbon compounds, the remaining oil becomes more dense and heavy. Tar balls and floating mats of heavy hydrocarbons may form.

Waves mix oil and water, dispersing oil droplets below the water surface. Over time, dispersion spreads oil compounds underwater even farther than the sheen of oil that can be seen on the surface.

In mixing oil and water, waves and winds often cause water droplets to become trapped within oil. This sticky, pudding-like mixture is called an emulsion. Water-in-oil emulsions are brown-colored and are called chocolate mousse, or just mousse. This emulsion slows the weathering process. Mousse may persist for as long as two years after an oil spill.

Sometimes oil and water do *mix, forming an emulsion called mousse.*

Sunlight also affects spilled oil in a process called photooxidation. Solar energy causes some petroleum compounds to mix with oxygen and form such compounds as peroxides, ketones, and alcohols. These substances are more likely to dissolve in water, and to harm living things, than their "parent" hydrocarbons.

Usually, some spilled oil settles to the ocean bottom, or is washed onto beaches. As an oil spill weathers and loses more and more of its lightest hydrocarbons, the remaining oil tends to sink. When ocean waters contain bits of soil, especially clay, the particles of earth stick to oil droplets. The more sediments that are mixed in the water, the more oil that is eventually deposited on the ocean bottom. Along beaches, storms also may carry oil-laden sand offshore and deposit it on the bottom.

Throughout the weathering process, bacteria and fungi play an important role in changing hydrocarbon compounds and removing them from the environment. This biodegradation of oil, as it is called, begins within hours after oil is spilled. The process of breaking down hydrocarbons by microscopic organisms (microbes) goes on at the surface, underwater, on beaches, and in sediments on the ocean bottom. It works best in warm waters that contain abundant oxygen, as well as such nutrients as nitrogen and phosphorus.

Under the best of circumstances, however, biodegradation is a slow process. Microbes have a difficult time breaking down tar balls and deposits of oil on the ocean bottom or along shores.

Oil lasts longest in protected coves and marshes, where there is little or no wave action. In contrast, along "high-energy" shores—where powerful waves scour the rocks and sand—the weathering process is speeded up.

Scientists learned a lot about the oil weathering process after the 1978 *Amoco Cadiz* spill off the coast of Brittany, France. Not only did the huge loss of 223,000 tons of crude oil attract international scientific interest, but the accident also occurred near several ocean research laboratories. Soon after the accident, scientists began collecting water and oil samples. *Amoco Cadiz* remains one of the best-studied oil spills in history.

Study of the Amoco Cadiz *wreck off the French coast led to a better understanding of the oil weathering process.*

High waves whipped some of the *Amoco Cadiz*'s cargo into a frothy mousse. Of the oil spilled, 30 percent evaporated and about 14 percent dispersed into the water. Another 4 percent was consumed in the open sea by microbes.

Eight percent of the oil was deposited in sediments on the ocean bottom, and 28 percent washed into the intertidal zone of the shore.

Oil fouled nearly two hundred miles of Brittany's coast, including sandy beaches, rocky coves, and salt marshes.

This disaster occurred in March, migration time for many birds. More than 3,200 dead birds of thirty species were recovered; scientists believed that this was only 10 to 15 percent of the real kill. High concentrations of oil were found two feet underground in some tidal areas. The oil caused massive kills of clams, sea urchins, crabs, and other marine life. Brittany's oyster-raising industry was harmed, and its fishing and tourism business suffered.

Within a year, however, catches of crabs, lobsters, and most fishes returned to normal. In one bay, three years passed before oil pollution lessened enough for oysters to survive. Microbes and weathering continued to rid habitats of oil. Nevertheless, a decade after the 1978 spill some heavily-oiled marshes still had harmful levels of hydrocarbons.

Although oil tankers have broken apart and even exploded in the open seas, most oil spills occur closer to shore. On any given day, about fifty tankers carrying 450 million gallons of oil enter American harbors. Coastal oil spills can be a threat to beach resorts and commercial fishing. They may foul salt marshes and estuaries, which are nurseries for young ocean fish.

Even if a spill fails to wash ashore, petroleum released anywhere over continental shelves can do harm. Compared with the open sea, water over the shelves, which may extend several hundred miles from continental shores, is shallow. It is also rich with the nutrients needed by the tiny plankton that are the base of many food chains. The upper few feet of these waters are home to a wealth of life, including fish eggs, fish larvae, and also crab and lobster larvae. And this precious layer, the ocean's skin, is exactly the area over which an oil spill spreads.

Oil may do long-lasting harm in salt marshes (left), where it breaks down slowly, in contrast to its effects along wave-swept shores (right).

The effects of petroleum on living things has been studied at the sites of spills and in laboratories. We know more about oil's effects on saltwater life than on freshwater organisms, because the biggest oil spills occur at sea, not on rivers or lakes. In either environment, hydrocarbons can kill plants and animals. Low doses can have "sublethal" effects; for example, not killing a fish directly, but harming its vision, its growth, or its ability to reproduce. A very low concentration of oil (five parts of oil in a million parts of water) affects fishes' sense of smell, and so impairs their ability to hunt.

Small-scale laboratory experiments may underestimate the harm done by oil pollution. Several of such studies, for example, showed little damage to coral reefs by oil. Then, in 1986, more than two million gallons of crude oil spilled from a refinery storage tank into the sea on the Carribean coast of Panama. The area was already well-known to scientists at the nearby Smithsonian Tropical Research Institute, and they began studying the oil's effects soon after the spill.

Low concentrations of oil impair the senses of fish and other aquatic creatures.

A 1986 oil spill off Panama did severe and long-lasting harm to coral reefs and the life that thrives on them.

The oil affected about fifty miles of shoreline habitats, seriously damaging the plant and animal life of mangrove forests, seagrass beds, and coral reefs. Some reefs were killed outright. Others suffered from sublethal effects, as algae and other organisms invaded their injured skeletons. Five years after the spill, just five percent of the reef was alive, and it had little chance of withstanding the eroding action of the sea.

Many forms of life are exposed to small amounts of oil in their natural environment. They usually appear unharmed, but may be affected by petroleum's sublethal effects. Chronic, low-level exposure to hydrocarbons occurs in many harbors and other coastal habitats where small oil spills happen. It occurs in estuaries, where oil carried by river waters mixes with sea water. It occurs near natural seeps of oil from cracks in the ocean floor.

Chronic oil pollution also occurs in the Persian Gulf. Before the Desert Storm conflict of 1991, the Gulf was considered the world's most oil-polluted body of water. It is a shallow sea, with only a narrow outlet, and pollutants tend to be trapped in its basin. However, the warmth of its water speeds the evaporation of oil, and its breakdown by microbes. The buoyancy of its highly salty water also tends to keep oil from sinking to the bottom.

Despite chronic oil pollution in its waters, the Persian Gulf supported a thriving fish and shrimp industry, and an abundance of wildlife, including dolphins, whales, sea turtles, dugongs (relatives of manatees), and many species of seabirds. All of this life existed despite oil spills that average at least a million barrels annually.

Iraq's deliberate oil spill in 1991 totaled an estimated six million barrels. It was the largest oil spill in history. The oil threatened life in the western Gulf, where the water is most shallow and

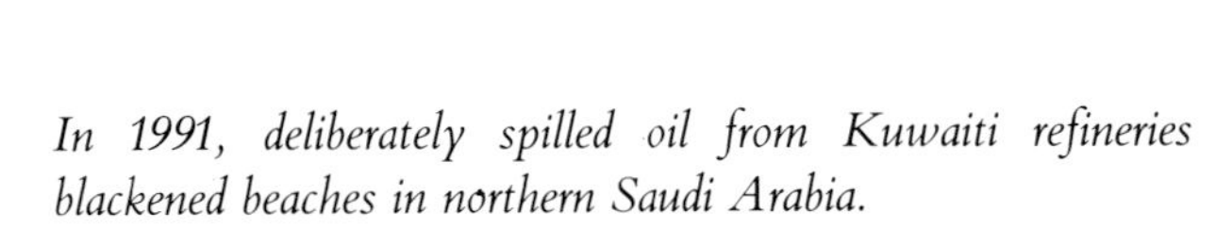

In 1991, deliberately spilled oil from Kuwaiti refineries blackened beaches in northern Saudi Arabia.

marine life is most abundant. Fish, algae, sea grass, mangroves, shrimp, and many other plants and animals that had survived long exposure to low levels of hydrocarbons died in great numbers.

In 1992, a team of scientists sponsored by the United Nations began assessing the damage. In Saudi Arabia, they found more than a hundred miles of beaches and shallow tidal zones covered by thick mats of tar. They found dead and dying coral reefs. In some areas of deep water, however, they found little apparent damage from the oil.

The scientists looked for signs that the life of the Persian Gulf could recover. "Some of the damage is permanent," said Dr. Sylvia Earle, then chief scientist of the National Oceanic and Atmospheric Administration (NOAA). "Life in the Gulf will recover, but it will be different."

Overall, wildlife populations and their habitats eventually do recover

A Persian Gulf cormorant, oil-covered and doomed.

from oil spills. Each spill is unique, however, and some pose a threat to endangered species. Oil from the Ixtoc-I well in the Gulf of Mexico moved toward the coasts of Mexico and Texas. Along the Texas coast there are seven national wildlife refuges, including the Aransas refuge, which is the winter home of the rare whooping crane. Oil did blacken some Texas beaches, but did not reach any of the major refuge areas.

Ixtoc-I's greatest threat was to the Atlantic Ridley sea turtle, which lays its eggs on beaches near Rancho Nuevo, Mexico. Baby turtles hatch in the summer, then swim northwest in the Gulf. In order to protect a generation of this endangered species, biologists caught about nine thousand hatchlings and airlifted them to the open sea beyond the oil slick.

Ever since the 1969 oil-well blowout off Santa Barbara, wildlife biologists and environmentalists have worried about the impact of an oil spill on seals

Oil from the Ixtoc-I well washed up on beaches of San Padre Island, Texas.

and sea otters. This concern has grown since 1977, when oil fields were first tapped in Alaska and tankers began carrying crude petroleum along the West Coast of the United States and Canada.

Some aquatic mammals, including whales, do not seem to be harmed by swimming through an oil slick. However, the same oil may doom aquatic mammals with fur, including sea otters, river otters, and fur seals. Oil on their fur reduces its insulating qualities. Sea otters, in particular, depend on their fur to keep warm. Their fur must be groomed frequently in order to be effective. Grooming is a key to their survival, but it can also seal their doom. In the process of licking their fur, they can swallow a lethal dose of oil. Badly oiled fur drives sea otters into a frenzy of grooming, which only makes matters worse.

Oil spills in several places have killed sea otters, and, along a coast in Scotland, river otters. Autopsies showed that they had died from damage to their livers, kidneys, lungs, or nervous systems. Since sea otters spend so much time feeding, grooming, and resting on the water surface, they may inhale benzene and other toxic hydrocarbons that usually evaporate soon after an oil spill. Sea otters have also died from exposure after their fur lost its insulating quality, and perhaps from eating oil-contaminated clams and other shellfish. These mollusks filter petroleum from the sea and cannot excrete it. They store hydrocarbons, and for weeks or months after an oil spill are a lingering threat to any animal that eats them.

The 1989 *Exxon Valdez* oil spill killed more than a thousand sea otters. These were known deaths; biologists believe that many others sank to the bottom without a trace. Before the accident, however, the otter population of Prince William Sound was estimated to be between twelve thousand and fifteen thousand. In time, barring further massive oil spills, the otter population will recover. Nevertheless, the *Exxon Valdez* spill showed how deadly an oil spill can be for sea otters. A spill of similar dimensions occurring along the California

Sea otters lick their fur frequently, so are especially vulnerable to oil spills.

coast might wipe out a separate, much smaller sea otter population entirely.

The full extent of the harm done by the *Exxon Valdez* oil spill may never be known. Years after the first massive loss of seabirds, bald eagles, sea otters, and other life, scientists continued to study the area.

They feared that the damage would be long-lasting. Oil had coated nearly twelve hundred miles of coastline, and many square miles of bottom sediments. (Had the spill occurred on the eastern coast of the United States, it would have stretched from Cape Cod, Massachusetts, to North Carolina.) In a warm climate, most of this oil would break down quickly, except in sheltered coves where waves were small. However, oil weathering is a slow process in

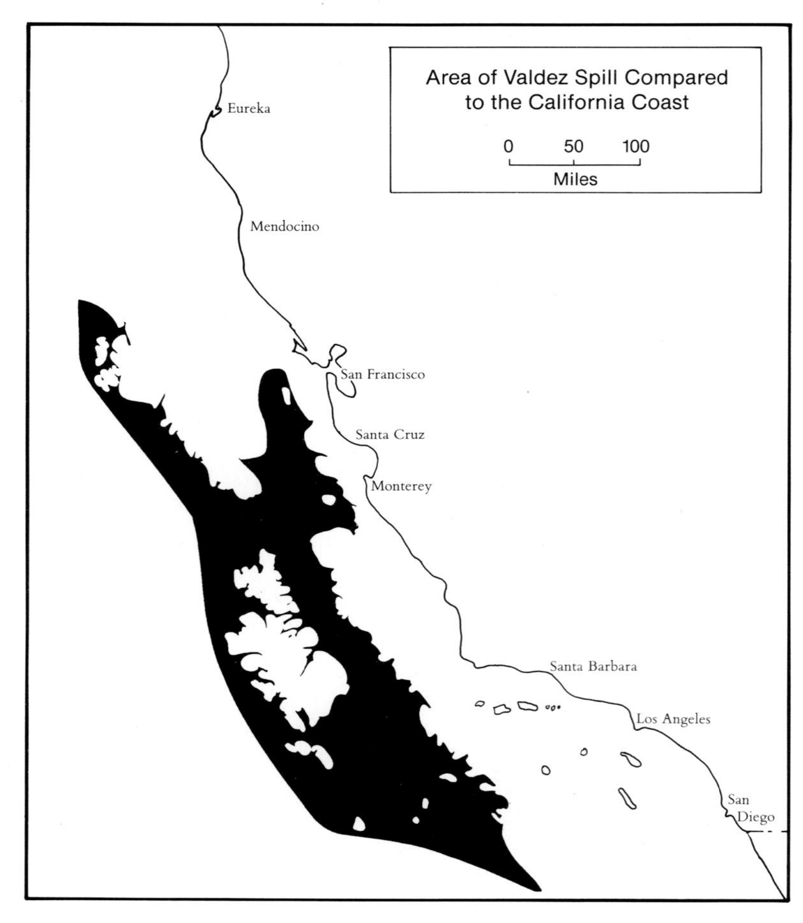

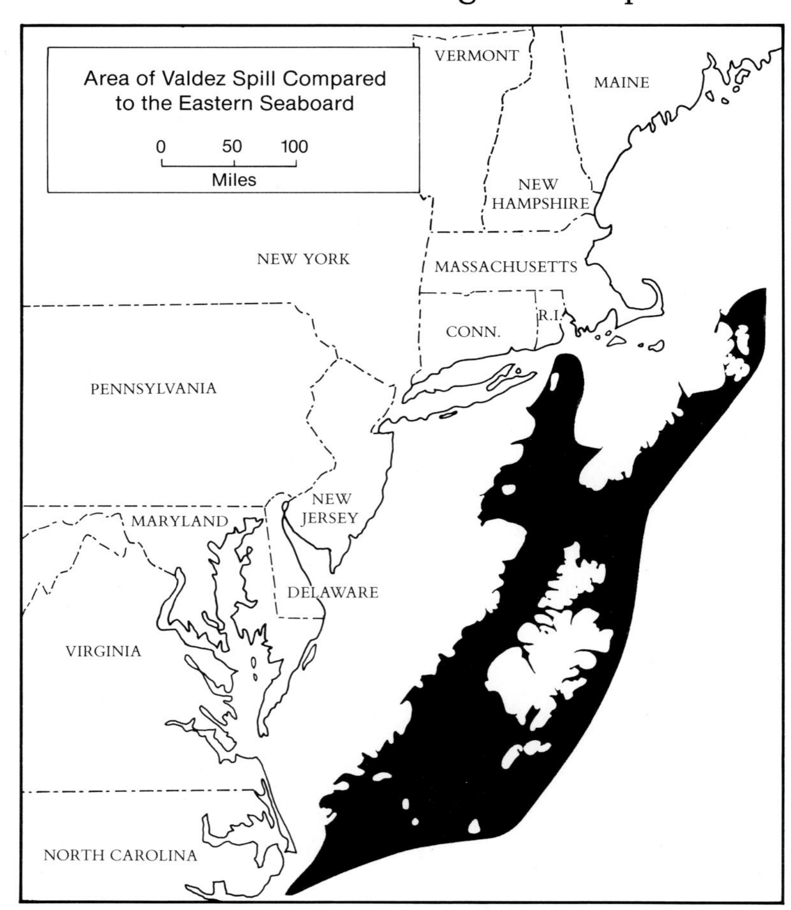

the subarctic climate of Prince William Sound. In limited sunlight and cold water, microbes break down oil molecules very slowly.

Two years after the spill, the United States government announced the results of some studies. Scientists had found many abnormal young herrings, and few or no salmon eggs and young in some salmon-breeding streams that flow into the sound. Some seabird colonies had lost nearly three-quarters of their populations. These colonies were located on islands where oil still persisted, and scientists feared that the birds could be completely wiped out.

Government scientists said that environmental damage at Prince William Sound would last much longer than they had originally thought. Some scientists predicted that recovery would take twenty or more years. In response, a spokesman for the Exxon Corporation, owner of the *Exxon Valdez* and its cargo, said, "Everything we are seeing suggests the biological community is healthy and thriving."

Some oil-doused bald eagles were successfully cleaned and released back into the wild.

4
Cleaning Up

The ugly black mess of an oil spill upsets people. They want something done about it, especially to protect beaches, fish, and wildlife. With a large petroleum spill, however, even a great cleanup effort falls short. Sometimes, the remedies do more harm than good.

If cleanup crews act quickly and a spill is spreading slowly, some of the petroleum can be recovered. One method is to surround the oil with inflatable, flexible, floating tubes called booms. These floating barriers work best in calm seas. Moderate currents and waves sweep oil under or over booms. The oil trapped by booms can be sucked through hoses into storage tanks on boats or barges. Also, boats equipped with mechanical skimmers can recover some oil.

Materials that soak up oil, but not water, are also used. They are called

The Exxon Valdez *was encircled with flexible booms in an attempt to prevent further loss of oil.*

absorbents, or just sorbents. These materials include straw, sawdust, minerals, and synthetic chemicals. After being spread over the surface of an oil slick, sorbents can be collected and the oil squeezed from them.

Methods like these recovered 1.5 million gallons of oil from the Persian Gulf in 1991—perhaps twenty percent of the amount spilled. In Alaska, eleven thousand workers may have recovered fifteen percent of the *Exxon Valdez* spill—at a cost of more than a billion dollars. Usually, no more than ten percent of a major oil spill is recovered.

All too often, the effort to contain spilled petroleum is late. Then booms may be used as barriers to keep oil from reaching shore. In 1991, booms were placed near islands in the Persian Gulf in order to keep oil from reaching the water around their shores—the habitat of birds, dugongs, and rare sea turtles.

Often, cleanup crews find that an oil spill (or parts of it) is spread too thinly over the water's surface to be recovered. Rather than try to contain it, they do the opposite: try to disperse it. From boats or aircraft, they spray chemicals called dispersants. These substances break up the oil into fine droplets and speed the dispersion process of oil weathering. Dispersants are commonly used on small oil spills. They can be an effective way of spreading petroleum through deep water and keeping it from reaching shore. The oil vanishes from sight, but is still present in the environment, and may harm life underwater.

Dispersants themselves can harm aquatic life. Laboratory studies have shown that some of these chemicals are more poisonous to fish than crude oil. Mixed with oil, the dispersants become even more toxic. These lab studies were prompted by the 1967 *Torrey Canyon* oil spill off the western tip of England. Aiming to disperse the huge oil slick, the Royal Navy dumped many thousands of gallons of detergents into the sea. The chemicals helped mousse to form,

Workers spread blotter-like materials to help clean up a small oil spill.

and caused a great loss of plant and animal life near and on the shore.

Ever since then, the use of dispersants has been controversial. Some states forbid their use. Furthermore, the wreck of the *Torrey Canyon* was only the first example of cleanup efforts causing harm. In Brittany, France, workers removed plants and even a layer of soil from a salt marsh that had been heavily dosed with oil from the *Amoco Cadiz*. These efforts to remove oil, along with the harm done by people trampling through the marsh, actually delayed the recovery of the marsh habitat. Ten years after the spill, heavily-oiled marshes that had been left alone were back to normal. The marsh that had been "helped" was not.

After the *Exxon Valdez* oil spill, there were also charges that some cleanup efforts had done more harm than good. Oil had come ashore on Alaskan beaches that are among the hardest to clean: those covered with pebbles or rocks, with gravel or sand underneath. On such shores, oil seeps down through spaces between rocks and into sediments that are protected from wave action.

Part of the massive and expensive cleanup effort involved workers actually washing beach stones one by one. Cleanup crews also sprayed rocky beaches with high-pressure hoses, trying to flush oil offshore, where it could be recovered. In the process, however, oil was also driven deeper into the beach sediments. A U.S. Coast Guard officer inspected a quarter-mile of beach where two hundred workers had spent eight days cleaning oil off rocks. He dug eight inches into the sediments and came up with both hands full of oil.

Whenever there is a large oil spill, the main public concern is the harm done to wildlife, especially birds and mammals. Within three days of the Exxon

In Alaska, water was sprayed on some rocky beaches in an attempt to wash the oil into Prince William Sound where it could be recovered.

Valdez accident, a major effort was launched to capture, clean, and release oiled birds and sea otters. Some biologists of the U.S. Fish and Wildlife Service questioned the usefulness of this rescue project. But the Exxon Corporation, seeking to clean its spill-blackened image, wanted an all-out animal rescue.

The attention of the public and news media was focused on sea otters. At its peak, the otter rescue effort employed 320 people at four treatment centers. Thirteen boats and a helicopter were used to capture otters. The six-month effort cost more than $18 million.

In all, 357 sea otters were captured. More than a third died in captivity. Some were given to zoos and aquariums. Some of the otters treated and released were equipped with radio transmitters so that their progress could be followed. Within eight months, half of these animals were dead or missing and presumed dead—a death toll that suggests that many other released otters also perished. Assuming the best for the released otters, the number of captured otters that survived in the wild or in zoos was 222. The cost for each of these rescued otters was more than $80,000.

Several wildlife biologists questioned the wisdom of the entire effort. They had observed that some captured otters were free of oil, or so lightly covered with oil that they probably had a good chance of survival in the wild. A few unoiled otters were actually chased into oil slicks by workers trying to catch them.

Biologists knew that many sea otters in Prince William Sound were not threatened by the oil spill. The entire otter population was not at risk, so the rescue was a humane effort to save as many oiled animals as possible. As one biologist put it: "We were doing it to save the animals that were caught in the slick—animals that were suffering, and animals the American public was demanding that something be done to help."

The effort, however, was ineffective and extremely expensive. "Should the time, money, and anguish be put forth to save a few individuals?" asked James

A. Estes, biology professor at the University of Santa Cruz, California. "We did it for the people," he said, "not for the animals. As a long-term strategy for conservation, we need to look toward prevention of oil spills, rather than cures."

Many sea otters died after being coated with oil from the Exxon Valdez. *Even those captured and cleaned had a poor survival rate.*

A major oil spill does offer opportunities to learn about "cures"—ways to clean up oil. In the summer of 1989, Exxon researchers tested a new method for cleaning oil from beaches. They sprayed seventy miles of beaches around Prince William Sound with a nitrogen-phosphorus fertilizer. The goal was to stimulate the growth of naturally occurring bacteria that feed on hydrocarbons. Some of the results seemed encouraging. Compared with untreated shores, some sprayed beaches showed dramatic improvement. Others did not, however, so more research on this technique is needed.

As long as oil is spilled, there will be a continuing need for better ways to clean it up. This is a business opportunity. It is being pursued by companies that are developing chemicals that help recover spilled oil before it spreads and disperses. New sorbent materials are being tested, as are chemicals that may help recover more oil.

Fertilizers sprayed on some Alaskan beaches seemed to help microbes break down spilled oil, but had no effect on other shores.

A Virginia company produces a substance called Elastol that acts in a manner opposite to a dispersant. Sprayed on an oil slick, it dissolves in the oil and makes it stickier. The oil is pulled into a compact mass that can be removed with suction equipment. (Applying too much Elastol makes the oil too thick to pump.) In 1990, Elastol was used effectively on a spill of heating oil in the harbor of New Haven, Connecticut. It kept the oil from spreading. The oil was then pumped back into the barge that had spilled it.

New remedies for spilled oil will be most effective with small spills, which are the most common by far. For any spill, however, the most effective cleanup force is nature itself. Given enough time, the weathering process, including microbes that feed on hydrocarbons, can remove every bit of petroleum. But the time varies, from a few months for a spill of gasoline, to years—sometimes many years—for crude oil.

As long as high concentrations of hydrocarbons remain in a marsh, beach, or other habitat, they may harm living things. So, relying on nature or on new ways to clean up spills is foolish. As biologist James Estes pointed out, there's a more important goal than simply cleaning up oily messes: stopping them from happening.

5

Preventing Oil Spills

In the wake of the Exxon *Valdez* accident, fourteen oil companies agreed to set up five centers in different regions of the United States that would be prepared to deal quickly with large oil spills. As usual, the oil industry's emphasis was on cleaning up. But many people were outraged over the massive oil spill in Alaska. The public had been assured that this was an accident that could not happen. It had, and people called for action that would prevent similar disasters in the future.

This led to the Oil Pollution Act of 1990, the latest of several United States laws and international agreements aimed at reducing petroleum spills. This law required the oil industry to make better preparations for the emergency of a spill. More important, it gave oil and shipping companies a strong financial reason to avoid spills. Unlike the previous system, which, according to a panel

Used oil is a resource that can be collected and processed into fuels rather than discarded to pollute our environment.

of scientists who studied the *Exxon Valdez* spill, provided "minimum penalties for creating massive environmental damage," the new law greatly raised the limits on liability for oil spill damages and cleanup costs. A spill resulting from carelessness can now cost a company hundreds of millions of dollars. The new regulations seemed to have had a good effect: The amount of oil spilled in 1991 was the lowest since 1978.

This law also set deadlines for United States oil tankers to have double bottoms, and eventually double hulls. Of the world's 3,200 oil tankers, about five hundred have double bottoms. Fewer have double hulls. In an accident, the bottom or sides of such tankers can be pierced but the second, inner bottom or hull helps prevent oil from spilling, or reduces the amount lost. According to the U.S. Coast Guard, had the *Exxon Valdez* been built with a double bottom, it would have lost twenty-five to sixty percent less oil. Adding these safety features to tankers will add to shipping costs, but will reduce the risk of major spills.

The Natural Resources Defense Council (NRDC) and other environmental groups worked hard for passage of the Oil Pollution Act of 1990. That accomplished, they urged other steps to reduce the risk of oil spills. One is to improve the systems that alert tankers to navigational hazards. Another is to strengthen the training and licensing requirements of the pilots who guide tankers in and out of harbors.

The NRDC also urged that the United States set up tanker-free zones in some areas. Tankers carrying oil from Alaska to the lower forty-eight states would be required to keep a hundred miles offshore. This would greatly reduce the chance of a major oil spill reaching the coast.

There is growing worry about oil spills because the demand for petroleum products grows each year. The United States is producing less and less oil. Early in the next century, as much as ninety percent of the oil consumed in the United States will arrive by tanker. Millions of tons of petroleum will be moving past vacation beaches, resorts, estuaries, fish spawning grounds, and fragile salt marshes.

Double hulls, pilot training, and higher penalties for oil spills should help to reduce tanker accidents. However, there are two other steps that would dramatically cut oil pollution. One is to set up a recycling program for used oil—the kind all too often dumped by do-it-yourself auto owners. Some communities have effective programs for collecting and recycling this oil, but a nationwide effort is needed to reduce this major source of pollution.

The second step is simple and touches everyone's life: *use less oil*. This would not only reduce tanker traffic and other sources of oil spills, but would stretch the use of precious petroleum farther into the future. One way to do this is to use more energy-efficient vehicles. In a sense, the United States's fleet of automobiles and light trucks is a huge, untapped oil field. Just improving engine efficiency by one mile a gallon would save 420,000 barrels of gasoline per day. Scrapping older, gas-guzzling cars would reduce petroleum demand much more.

Similarly, improvements in home and business heating systems—for example, installing highly efficient furnaces—could save as much oil as is produced in Alaska daily, about two million barrels.

Since the early 1970s, environmental groups and growing numbers of economists have urged that the United States adopt a national energy policy based on conservation. The U.S. government has no official energy policy, but unofficially it encourages extravagant use of gasoline and other petroleum products. Thanks in part to the politically powerful oil industry and automobile manufacturers, the United States is the world's most car- and oil-dependent nation.

Japan and Germany offer dramatic proof that petroleum use can be cut sharply without harming economic growth. These nations use half as much energy as the United States to produce each dollar of gross national product. If the United States used energy as wisely, its annual bill for oil and other fuels would fall by an estimated $200 billion.

Some people have already taken steps toward a less wasteful lifestyle. When the entire United States cuts its use of petroleum, its citizens will save money. They will breathe cleaner air. And North America's rivers, bays, and beaches will suffer less from those double disasters, oil spills.

Glossary

absorbents—porous materials used in oil spill cleanups that tend to soak up oil and not water; also called sorbents. Some are synthetic substances, such as polypropylene. Others include straw, sawdust, and a fine powder made from volcanic rock.

biodegrade—to break down or decompose through the action of bacteria and other microorganisms.

cracking—the chemical process of breaking down the heavier molecules of petroleum into lighter, more desirable products, such as gasoline.

crude oil—petroleum in its natural state before it is refined; sometimes this oil is simply called "crude."

dispersant—a solvent or other chemical agent used in oil spill cleanups that breaks oil into tiny droplets, helping to disperse it through a large volume of water.

emulsion—a fluid in which one substance (for example, oil) is dispersed as small droplets within another (for example, water). An emulsion of oil in water is brown in color, and is called mousse, or even chocolate mousse, after the frothy dessert.

estuary—a place where salt water and fresh water mix; for example, where a river enters an ocean. Estuaries are rich with nutrients and life, and are vital nurseries for many kinds of ocean fish.

fossil fuels—fuels containing carbon that formed from living materials millions of years ago. Petroleum, natural gas, and coal are fossil fuels.

fractions—various hydrocarbon compounds that are refined from crude oil, including gasoline, heating oil, and jet fuel.

hydrocarbons—chemical compounds made of hydrogen and carbon molecules.

microbes—microscopic-size organisms, including bacteria and some fungi.

molecule—the smallest possible amount of a compound that still has the physical and chemical characteristics of that substance.

photooxidation—the process by which oil reacts with oxygen in the presence of sunlight, and is changed chemically into new substances.

plankton—tiny, drifting plants and animals that live in salt and fresh water. The plants, or phytoplankton, include diatoms and algae. The animals, or zooplankton, include rotifers and copepods.

weathering—ordinarily a term applied to the gradual breakdown of rocks into soil, weathering is also applied to the natural processes that gradually rid an environment of spilled petroleum. The processes include dispersion, photooxidation, and biodegradation.

Further Reading

Anderson, Madelyn. *Oil Spills*. New York: Franklin Watts, 1990.

Borelli, Peter. "Troubled Waters: Alaska's Rude Awakening to the Price of Oil Development." *Amicus Journal*, Summer 1989, pp. 11–20.

Canby, Thomas. "After the Storm." *National Geographic*, August 1991, pp. 2–33.

Carr, Terry. *Spill! The Story of the Exxon Valdez*. New York: Franklin Watts, 1991.

Committee on Effectiveness of Oil Spill Dispersants, National Research Council. *Using Oil Dispersants on the Sea*. Washington, D.C.: National Academy Press, 1989.

Davidson, Art. *In the Wake of the Exxon Valdez: The Devastating Impact of the Alaska Oil Spill*. San Francisco, CA: Sierra Club Books, 1990.

Estes, James. "Catastrophes and Conservation: Lessons from Sea Otters and the Exxon *Valdez*." *Science*, December 13, 1991, p. 1596.

Geraci, Joseph, and David St. Aubin, eds. *Sea Mammals and Oil: Confronting the Risks*. San Diego, CA: Academic Press, 1990.

Green, J. and M.W. Trett, eds. *The Fate and Effects of Oil on Freshwater*. New York: Elsevier Applied Science, 1989.

Gundlach, Erich, *et al.* "The Fate of *Amoco Cadiz* Oil." *Science*, July 8, 1983, pp. 122–29.

Hardy, John. "Where the Sea Meets the Sky." *Natural History*, May 1991, pp. 59–65.
Hodgson, Bryan. "Alaska's Big Spill: Can the Wilderness Heal?" *National Geographic*, January 1990, pp. 4–43.
Jackson, J. B. C., *et al*. "Ecological Effects of a Major Oil Spill on Panamanian Coastal Marine Communities." *Science*, January 6, 1989, pp. 37–44.
Luoma, Jon. "The Ills of Oil." *Wildlife Conservation*, November–December 1990, pp. 35–53.
Marshall, Eliot. "Valdez: The Predicted Oil Spill." *Science*, April 7, 1989, pp. 20–21.
Nardo, Don. *Oil Spills*. San Diego, CA: Lucent Books, 1990.
O'Donoghue, Brian. *Black Tides: The Alaska Oil Spill*. Anchorage, AK: Alaska Natural History Association, 1989.
"Positive Energy: First Steps Down a Safer, Softer Path." *Sierra*, March–April 1991, pp. 36–47.
Stone, Richard. "Oil-Cleanup Method Questioned." *Science*, July 17, 1992, pp. 320–21.

Index

Photographs are in **boldface**.